ALICE M. WARD
MEMORIAL LIBRARY
P.O. BOX 134
CANAAN, VT 05903

# SATURN

Robin Kerrod

Lerner Publications Company • Minneapolis

**This edition published in 2000**



Lerner Publications Company.
A Division of Lerner Publishing Group
241 First Avenue North, Minneapolis MN 55401

Website address: www.lernerbooks.com



**Library of Congress Cataloging-in-Publication Data**

Kerrod, Robin.
Saturn / Robin Kerrod.
cm.—(Planet library)
Includes index.
Summary: An introduction to the planet Saturn, with information about its atmosphere, rings, and moons.
ISBN 0-8225-3909-8 (lib. bdg.)
Saturn (Planet) Juvenile literature. [1. Saturn (Planet) 2. Planets.]
I. Title. II. Series; Kerrod, Robin. Planet library.
QB671.4.K33 2000 99-26242
523.46—dc21

Printed in Singapore by Tat Wei Printing Packaging Pte Ltd
Bound in the United States of America
1 2 3 4 5 6 – OS – 05 04 03 02 01 00

# Contents

# Introducing Saturn

Saturn is the second largest planet in the solar system—the family of bodies that circle in space around the Sun. Saturn's most unique feature is the bright set of rings that travel around it. Three other planets have rings, but only Saturn's rings can be seen from Earth. These rings make Saturn one of the most beautiful bodies in the solar system.

Saturn, along with Jupiter, Uranus, and Neptune, is one of the gas giants. Unlike Earth, Mercury, Mars, and Venus, which are made up mainly of rock, Saturn is made up mainly of gas. Its makeup causes it to be extremely light for its size. In fact, it is the lightest planet in the solar system. If you could put it in water, it would float. Every other planet would sink.

Like all the gas giants, Saturn travels through space with a large family of moons. Of all the planets, Saturn has the largest family of known moons—at least 18. One of them, Titan, is larger than the planet Mercury. Titan is also the only moon in the solar system that has a thick atmosphere, or a layer of gases around it.

Saturn's rings are actually made up of thousands of tiny ringlets.

# Saturn Basics

**Saturn is a rapidly spinning planet, made special by the rings that travel around it.**

Saturn is the sixth planet in the solar system in order going out from the Sun. On average, it lies about 877 million miles (1.4 billion km) away from the Sun. It is nearly 10 times farther from the Sun than Earth is. At this distance, the planet takes nearly 30 Earth-years to circle, or orbit, once around the Sun.

Because it is so remote, Saturn is not an easy planet to spot in the night sky. During most of the year, it does not shine as brightly as Venus, Jupiter, or Mars, which are all closer to the Sun. But at its brightest, Saturn can outshine most of the stars in the sky. Like all the planets, Saturn does not shine by its own light. It shines because it reflects light from the Sun.

Saturn is so large that it could swallow nearly 750 bodies the size of Earth.

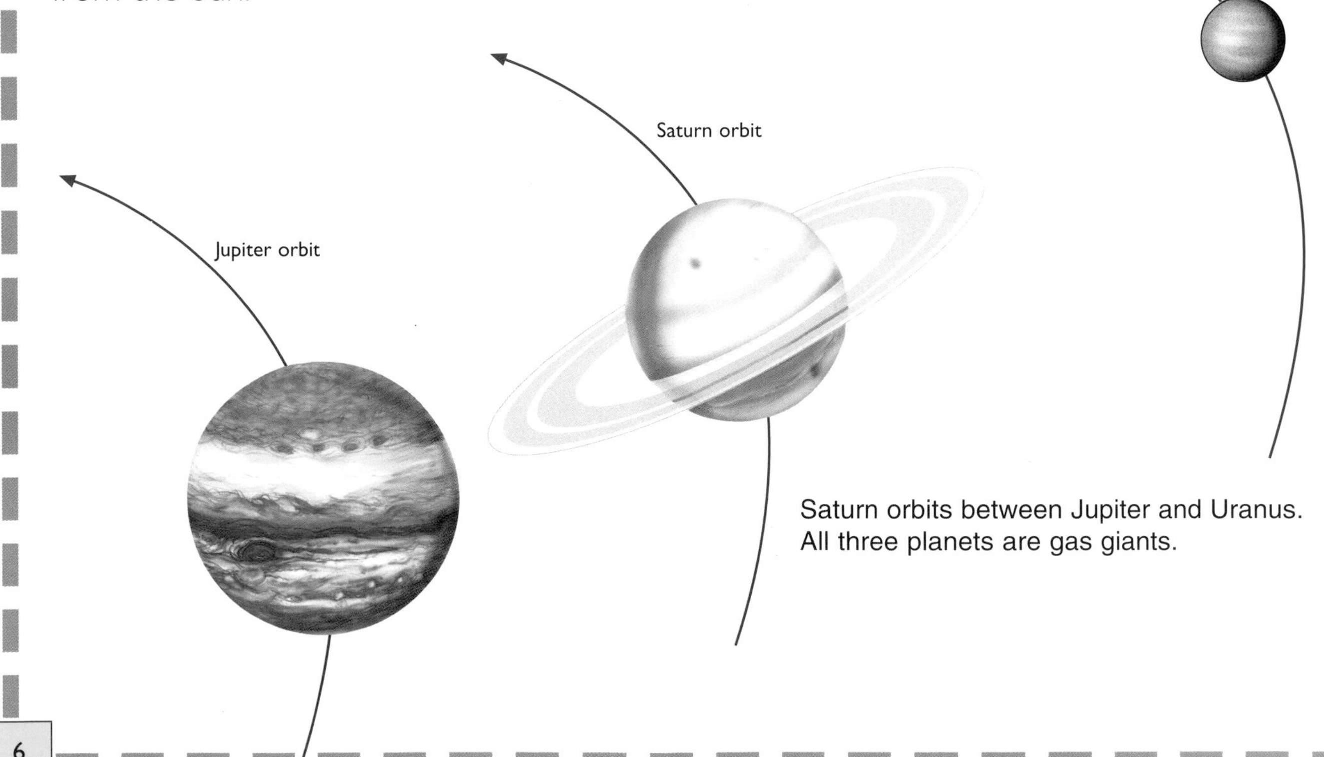


Saturn orbits between Jupiter and Uranus. All three planets are gas giants.

In this close-up view of Saturn, we can see how the planet casts a shadow on its rings.

### Through a Telescope

Through a small telescope, Saturn looks different from the other planets. The planet appears to be oval shaped rather than spherical, or round. A more powerful telescope will show why—the planet is surrounded by rings. The edges of the rings on either side make the image look oval.

Saturn appears to us to be yellow. This yellow coloring is not the planet's surface but its thick atmosphere. The main marking we can see on Saturn's yellow body is the black shadow cast by the rings. Occasionally we can spot small black dots, which are the shadows of Saturn's larger moons. Faint parallel bands can also be seen on Saturn. Astronomers call the darker bands belts and the lighter ones zones. Other faint dark and light streaks and spots appear on Saturn from time to time.

### How Saturn Formed

Saturn was formed along with the other planets about 4.6 billion years ago. It formed from gases and icy chunks of rock that came together in the outer solar system. Over time, the rocky materials formed into a larger and larger ball, which astronomers believe grew to become a body at least the size of Earth. Then the rocky body attracted gases around it until it became the gas giant we call Saturn.

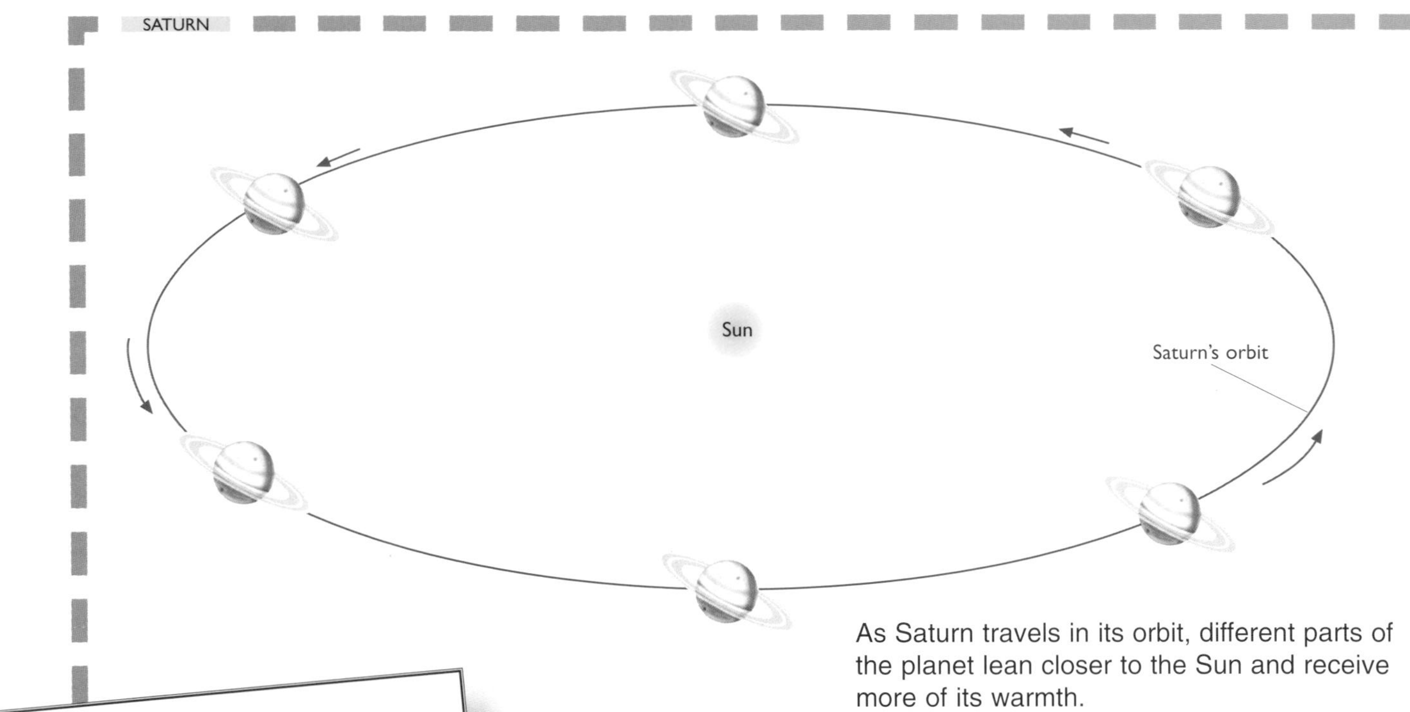


As Saturn travels in its orbit, different parts of the planet lean closer to the Sun and receive more of its warmth.

## Saturn's Warmth

Given Saturn's great distance from the Sun, temperatures on the planet are much higher than one would expect. Scientists believe this may be partially caused by droplets of liquid helium that form beneath the weight of Saturn's atmosphere. Because liquid helium is much heavier than liquid hydrogen, the droplets sink through Saturn's hydrogen ocean. As they fall, they give off heat. Over millions of years, this effect has caused temperatures on the planet to rise. Temperatures below Saturn's atmosphere may be at least 10,000°F (5,500°C), but its upper atmosphere is only about –300°F (–185°C).

### In a Spin

Saturn takes a little over 10½ hours to rotate, or spin around, once. This is less than half the time that Earth takes to complete one rotation, which is 24 hours, or one day. In fact, except for Jupiter, Saturn rotates more quickly than any other planet in the solar system.

Because Saturn rotates so quickly, it bulges out at the equator and is flattened at the poles. A planet's equator is the imaginary line around it, midway between its north and south poles. This flattening also occurs on Earth, but on Saturn, the flattening and bulging effects are much greater. That's because the planet is made up mainly of gas and liquid, which can change shape easily. This makes Saturn's diameter, or distance across, nearly 7,500 miles (12,000 km) greater at the equator than at the poles.

### Saturn's Tilt

Like all planets, Saturn rotates on its axis—an imaginary line running through a planet from its north pole to its south pole. In some planets, the axis is almost vertical, or upright, in relation to the planet's orbit. But in Saturn, the axis is tilted at an angle in space. As the planet travels in its orbit, this tilt causes different parts of the planet to lean closer to the Sun. The same thing occurs on Earth and brings about the regular changes in the weather we call the seasons. Because Saturn's orbit takes nearly 30 Earth-years, its seasons last for 7½ years.

Saturn's tilted axis also affects the way the planet appears from Earth. As the planet circles the Sun, we can see different views of the rings. When the planet is tilted toward us, we get our best views of the rings. At other times, the rings seem to almost disappear.

## Inside Saturn

Like its neighbor Jupiter, Saturn is made up mostly of gas. Unlike Earth, it does not have a solid surface that you could stand on. Beneath a heavy atmosphere of mostly hydrogen gas, great pressure turns the hydrogen into liquid. Astronomers believe that a vast liquid hydrogen ocean, tens of thousands of miles deep, covers the whole planet. Below this hydrogen ocean, even higher pressures turn the hydrogen into liquid metal. This liquid metal probably surrounds a small core of rock and ice at least the size of Earth.

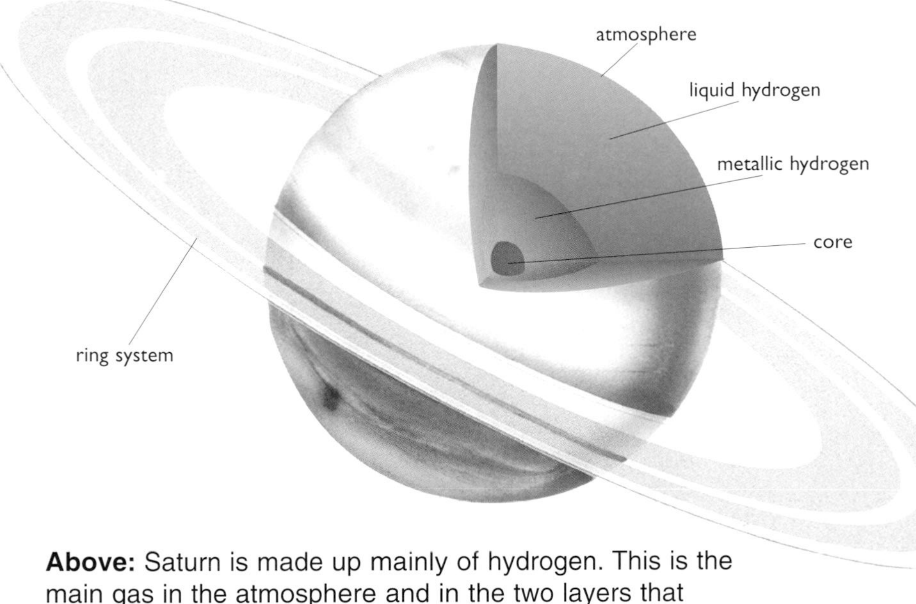


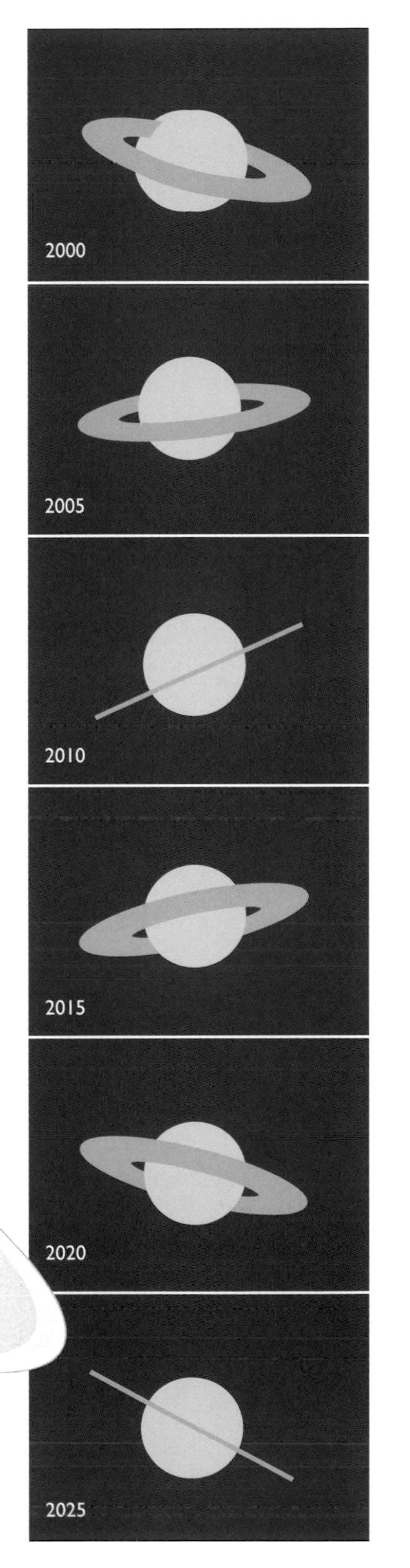


**Above:** Saturn is made up mainly of hydrogen. This is the main gas in the atmosphere and in the two layers that surround Saturn's core.

**Right:** Saturn's tilted position in space causes us to see different views of its rings from Earth. The diagram shows how our view of them will change over time.

**SATURN DATA**

**Diameter at equator:** 74,600 miles (120,000 km)

**Average distance from Sun:** 887,000,000 miles (1,430,000,000 km)

**Rotates in:** 10 hours, 39 minutes

**Orbits Sun in:** 29.5 years

**Moons:** 18 known

## Saturn's Magnetism

Like Earth and Jupiter, Saturn has a liquid metal layer that causes it to be magnetic. As Saturn rotates, the movement creates currents of electricity in its metal layer. These currents produce Saturn's magnetism. This is similar to the way Earth produces its magnetism. On Earth, magnetism is the force that makes a compass needle point north.

Saturn's magnetism extends many millions of miles out into space around the planet, forming a magnetic bubble known as the magnetosphere. A stream of particles from the Sun called the solar wind flows around the magnetosphere. Some of the particles in the solar wind get trapped in certain parts of Saturn's magnetosphere. The Earth's magnetosphere has similar areas, called the Van Allen belts, which produce powerful radiation. Other solar wind particles enter Saturn's atmosphere and make it glow. These light displays, or aurora, are similar to the Northern and Southern Lights we sometimes see on Earth.

Saturn's magnetism stretches far out into space. It acts as a barrier to the solar wind, and causes it to change direction.

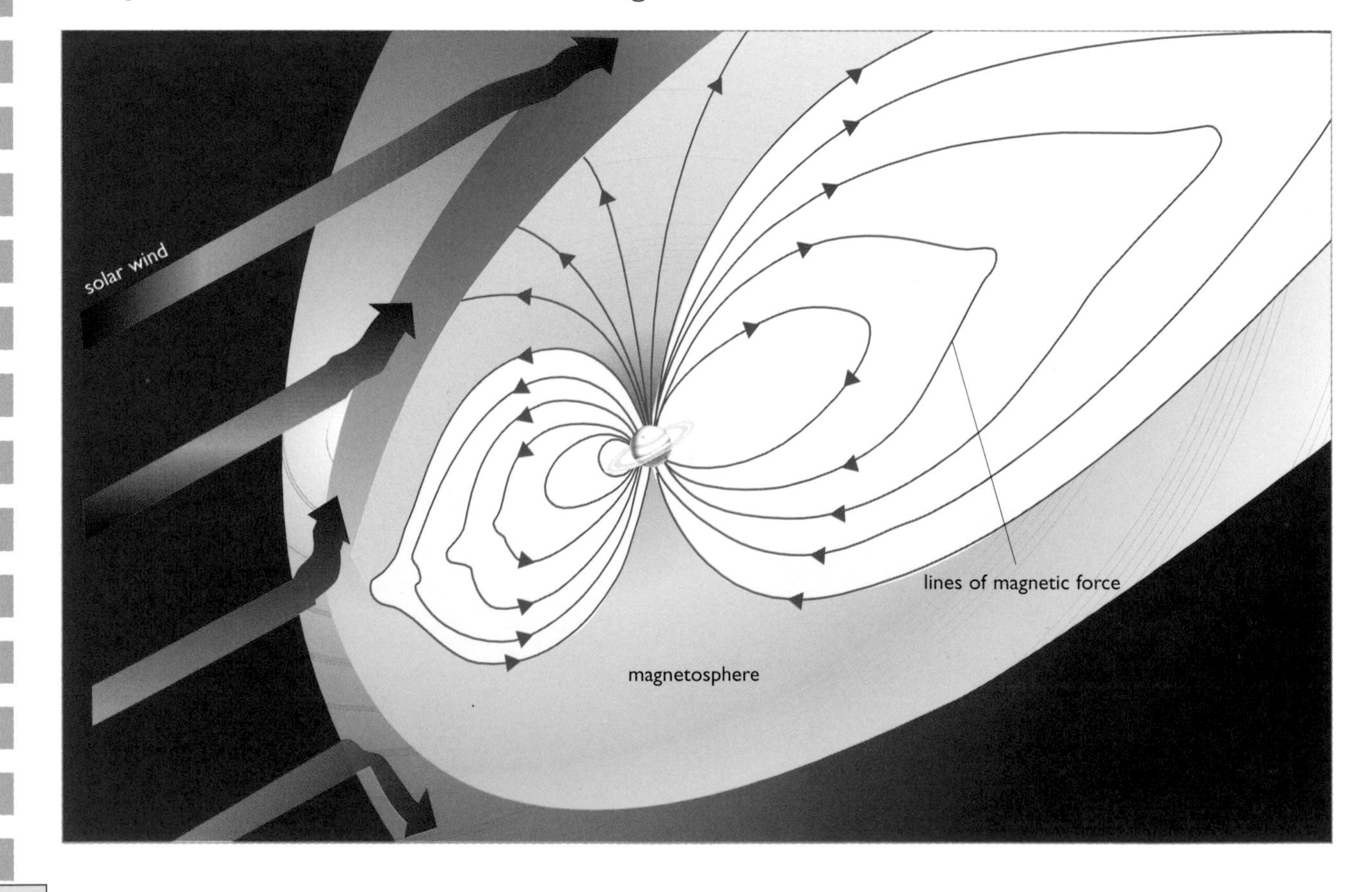

Particles from the solar wind that whiz around in Saturn's magnetosphere send out radio waves into space. Astronomers can pick up these waves, which make Saturn a kind of planetary radio station.

# Saturn's Atmosphere

**Strong winds blow in Saturn's atmosphere, causing clouds to swirl around the planet and creating fierce storms.**

Hydrogen makes up about 94 percent of Saturn's atmosphere, and helium makes up most of the rest. There are also traces of other gases, including methane, ammonia, and water vapor. These gases form the cloud features that appear as the faint bands called belts and zones that we see on Saturn. Partly hidden by a thick haze in the upper atmosphere, these clouds create Saturn's pale yellow color.

Temperatures in Saturn's upper atmosphere are very low, and the clouds there are made up of frozen gas crystals. Beneath these high clouds are clouds formed from water droplets. This lower layer of clouds is similar to clouds in Earth's atmosphere.

This colorized picture of Saturn shows the bands of clouds in the planet's atmosphere. Without colorization, the parallel bands are difficult to see.

## Wild Weather

In Saturn's atmosphere, winds blow violently, furious storms rage, and lightning flashes. On Earth, heat from the Sun produces our weather. But on Saturn, heat from inside the planet creates most of the planet's weather.

In this prominent band in Saturn's northern hemisphere, winds travel at speeds approaching 300 miles (500 km) an hour.

## Powerful Jets

Saturn's winds blow in powerful jet streams, or fast-moving air currents in the atmosphere. The most powerful one is the broad equatorial jet stream that extends for thousands of miles on either side of Saturn's equator. Winds within this jet stream whip around the planet at speeds of up to 1,100 miles (1,800 km) an hour. This is more than three times as strong as the winds in Earth's fiercest tornadoes.

The equatorial jet stream blows toward the east, in the same direction as Saturn's rotation. Farther north and south of the equator are several narrower jet streams that blow in the opposite direction.

The oval regions in this colorized picture of Saturn's atmosphere are violent storms.

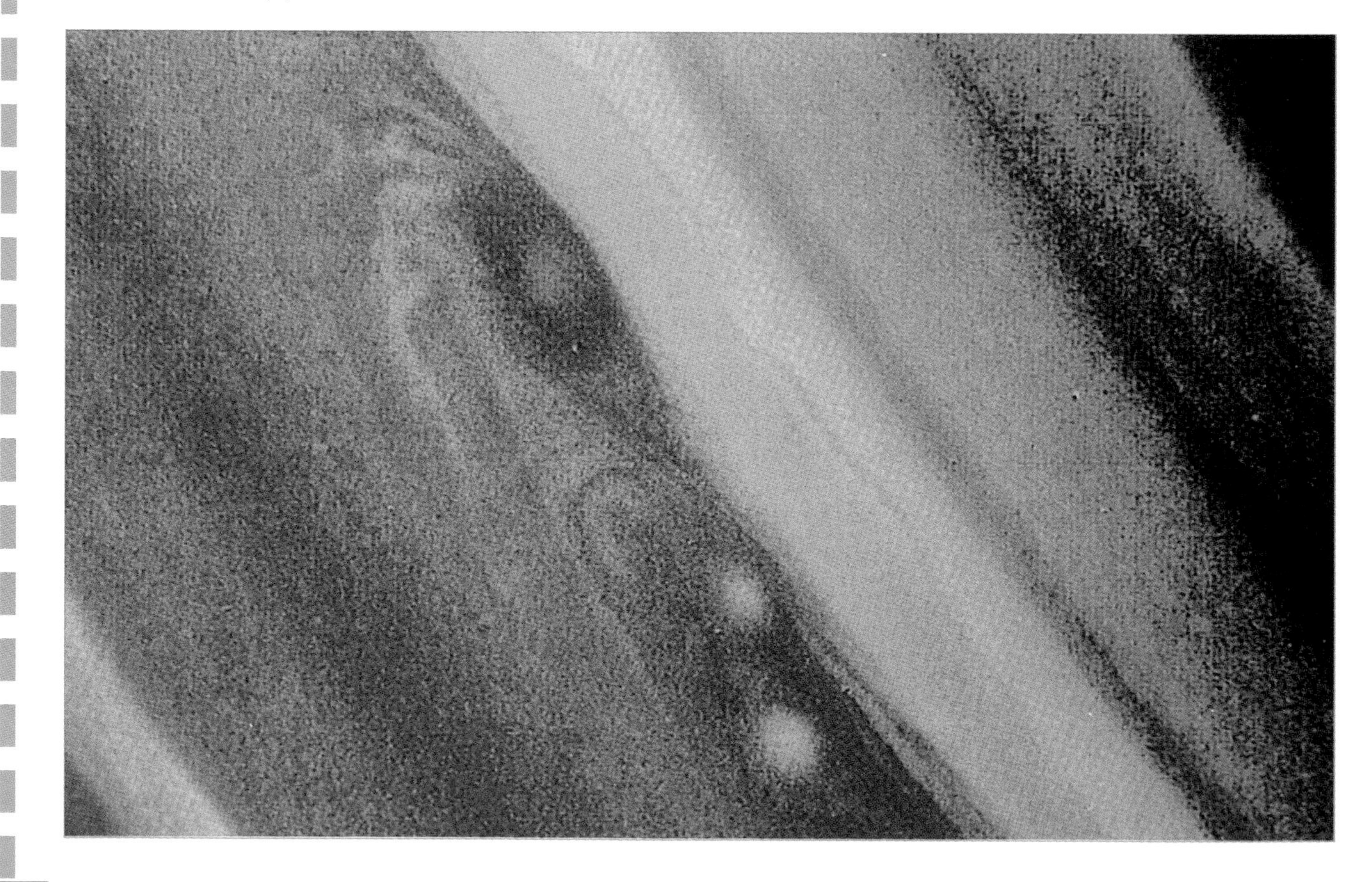

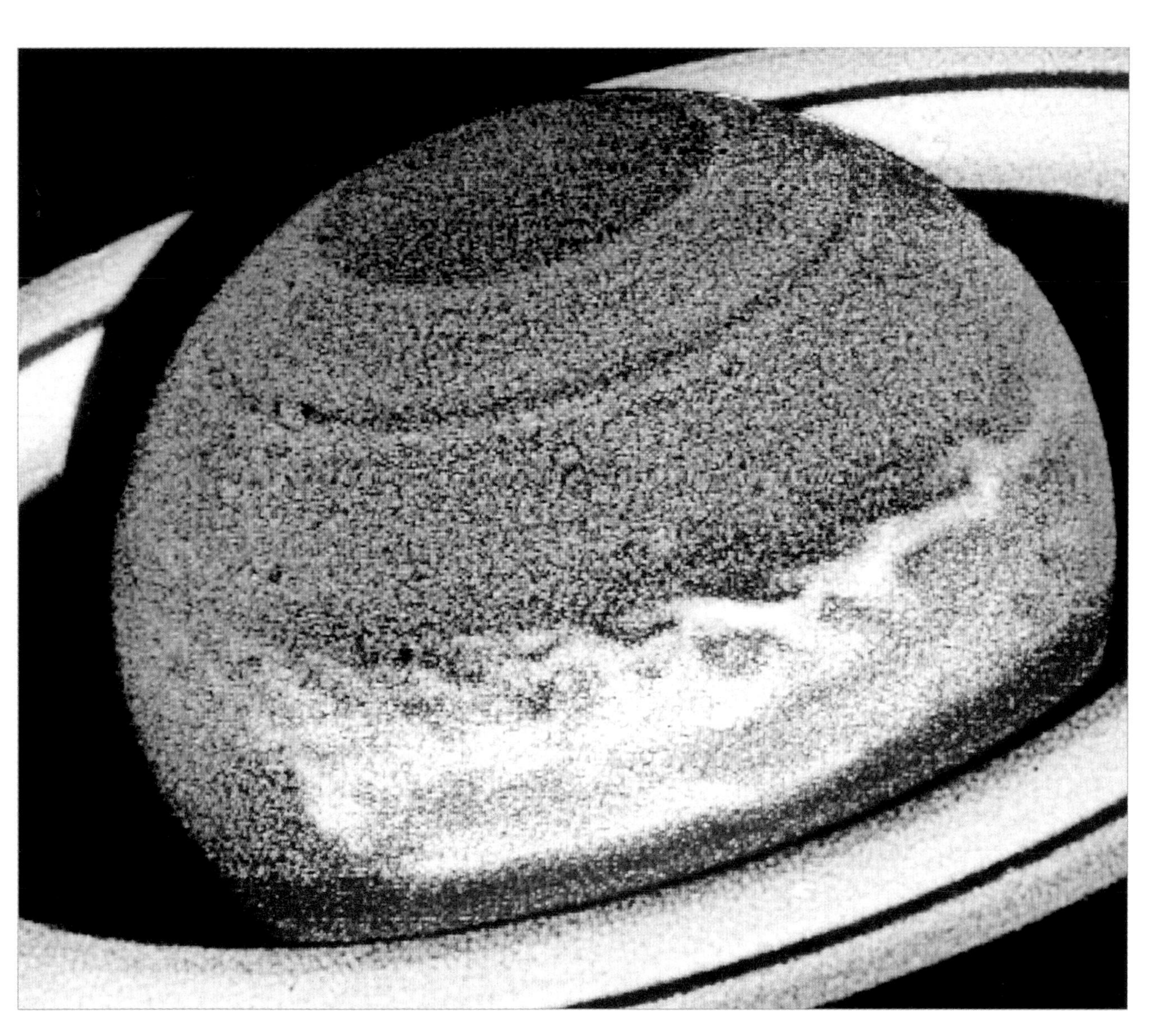

This Hubble Space Telescope picture shows Saturn streaked by violent weather and high winds that reach more than 1,100 miles (1,800 km) per hour.

## Spotting Storms

Violent storms occur at the edges of Saturn's jet streams, where the atmosphere gets churned up. Storms also occur within the jet streams themselves, caused by warm air circulating upward from lower levels in the atmosphere. From Earth, large storm areas in Saturn's atmosphere often appear as oval patches or spots. These spots can look white, brown, or red. They are similar to the spots that appear on Jupiter but are not as striking or as big. Nothing on Saturn is like the huge storm on Jupiter known as the Great Red Spot, which has been seen on the planet for centuries.

# The Remarkable Rings

**Four planets in the solar system have rings, but only Saturn's large rings are bright enough to be viewed through telescopes on Earth.**

When the Italian astronomer Galileo discovered Saturn in 1610, he noticed that it looked different from the other planets he had observed with his telescope. Other planets appeared to be round, but Saturn seemed to have ears, or strange attachments on each side. Galileo thought

**Above:** This superb picture of Saturn's rings shows the many ringlets that make up Saturn's main ring system.

**Below:** A photograph of Saturn's rings taken from directly above the planet's north pole

they might be moons. His telescope was not powerful enough to show what the "ears" really were—the outer edges of a huge system of rings.

Saturn's rings circle around the planet's equator. From edge to edge, the ring system measures more than 250,000 miles (400,000 km) wide. On Earth, we can see Saturn's three main rings. Astronomers have labeled them A, B, and C from the outside going in. From edge to edge, the A, B, and C rings appear to be about 40,000 miles (64,000 km) wide.

### Shining Bright

Saturn's rings reflect sunlight brilliantly. Without its rings, the planet would appear much fainter in the night sky. We know this because at two points in Saturn's 30-year journey around the Sun, the rings almost disappear from our view. When this happens, Saturn fades noticeably.

The three rings vary widely in their individual brightness. The middle B ring is the brightest, followed by the outer A ring. The inner C ring is very faint and transparent. The body of the planet can be seen through it.

This picture of saturn's rings is unusual because the normally bright B ring is not visible.

### More Rings

Along with Saturn's three major rings, astronomers have discovered several other rings. A faint ring, which scientists have labeled the D ring, circles close to Saturn inside the C ring. This ring is so close to Saturn that its inner edge probably touches the planet's atmosphere. More rings, which scientists have labeled F, G, and E, circle Saturn outside the A ring. The F and G rings are narrow, while the E ring is wide but very faint. This E ring is so far from Saturn that it travels between the orbits of Saturn's inner moons, Mimas and Enceladus.

Encke Division
A ring
Saturn
C ring
Cassini Division
B ring

**Above:** This diagram shows Saturn's main ring system, as seen from a point above the planet's north pole.

**Left:** Saturn's rings are actually made up of many small particles that circle the planet at high speed.

## Racing Ringlets

Astronomers have known for centuries that Saturn's rings could not be made up of one solid sheet of matter. Close-up pictures taken by space probes have shown that each ring is actually made up of hundreds or thousands of narrower ringlets. These ringlets are made of moving objects that vary in size from fine dust-like particles to chunks as big as trucks. Scientists think that the particles are mostly icy lumps and some ice-coated rocks. They look like rings because the light they reflect is blurred as they whiz around the planet.

## Ring Gaps

The rings we see from Earth do not form one continuous set of rings. In 1672, the Italian-French astronomer Giovanni Domenico Cassini spotted a gap between the A and B rings.

It became known as the Cassini Division. More than 150 years later, the German astronomer Johann Encke noticed a narrower gap inside the A ring, known as the Encke Division. From Earth, the Cassini and Encke Divisions appear to be empty gaps. However, they actually contain some ringlets. Many other smaller gaps exist where there are fewer ringlets than usual.

Scientists are not sure why these gaps exist. They may be caused by the gravity, or pull, of nearby moons, such as Mimas. This moon's gravity may pull particles away from certain parts of the rings.

**Above:** This close view of Saturn's A ring shows variations in its ringlets. The dark gap in the rings is the Encke Division.

## Where the Rings Came From

For many years, astronomers believed that Saturn's rings had formed from the remains of bodies that strayed close to the giant planet. If a smaller body, such as a moon, traveled close to Saturn, the massive planet's gravity could have eventually pulled the body to pieces.

However, the more we learn about the rings, the more mysterious they become. Astronomers are no longer certain exactly where Saturn's large ring system came from. Some scientists believe that the rings we see cannot be the remains of smaller bodies broken up and scattered long ago. Instead, Saturn's ring system may be continually forming and reforming over time.

### The Spokes

Saturn's rings have an interesting feature that cannot be seen from Earth. Astronomers have discovered that the B ring has spokes. These markings are dark lines fanning out across the B ring, similar to the spokes of a wheel. Up to thousands of miles long, they often appear and disappear within a few hours. Astronomers think the spokes might be caused by fine dust particles that lift up above the rings.

Saturn's rings are wide but very thin. In certain parts, they are probably less than 100 feet (30 m) thick from top to bottom.

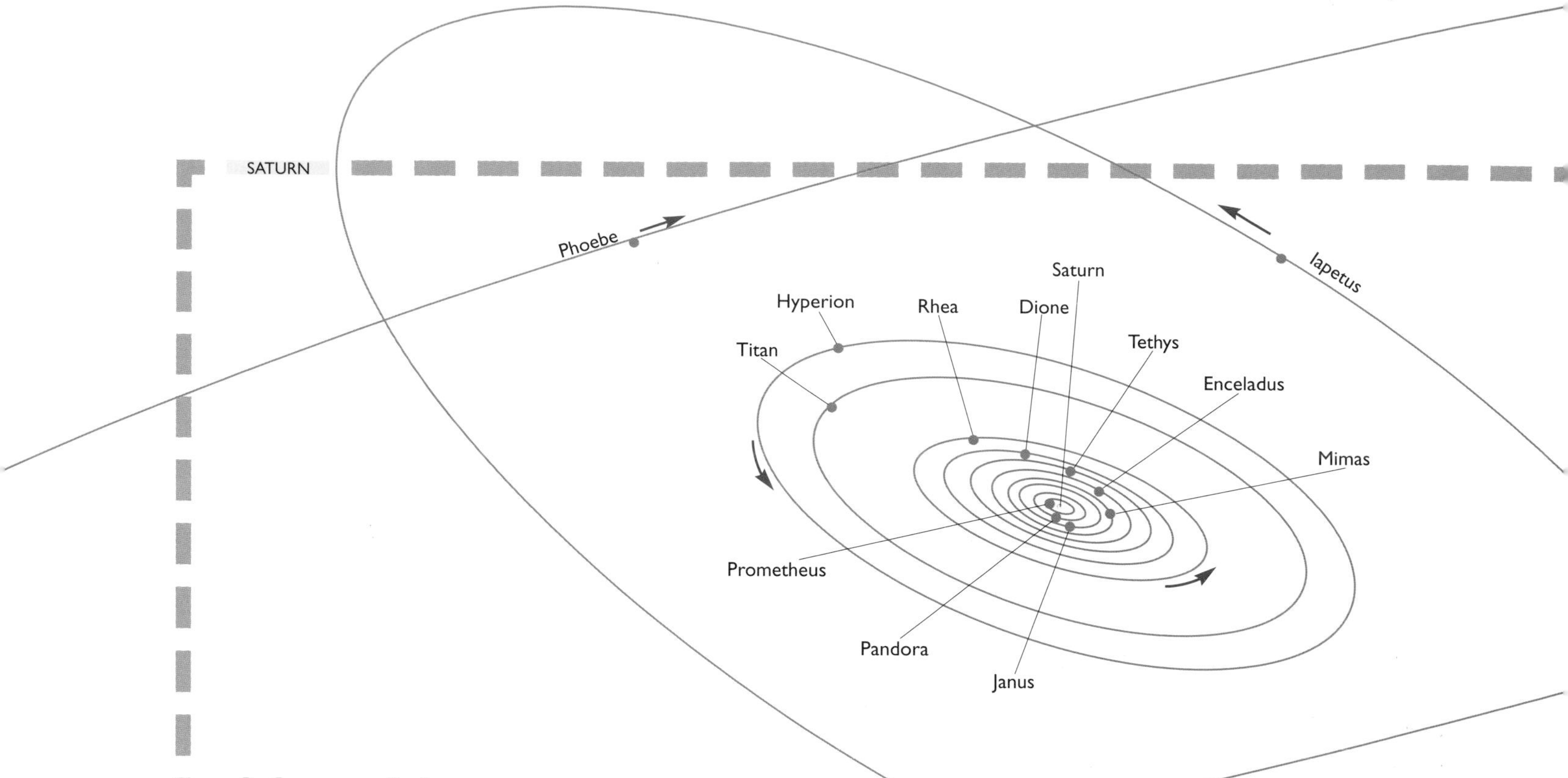

# Many Moons

**Saturn's 18 known moons vary in size and shape, from rocky lumps just a few tens of miles across to round bodies the size of planets.**

Most of Saturn's moons lie relatively close to the planet, but several moons orbit millions of miles away.

Before space travel was possible, astronomers believed that Saturn had only 11 moons—the ones they could see through telescopes. But space probes allowed astronomers to discover that the planet has at least 18 moons. Other possible moons have been reported, so Saturn may have more moons.

Most of Saturn's moons lie relatively close to the planet. Fourteen moons orbit within a distance of about 330,000 miles (530,000 km) from the planet. They form a kind of inner moon system. Several other moons orbit beyond a vast gap in the moon system, at great distances from the planet.

Like our own Moon, most of Saturn's moons complete one rotation in the same amount of time they take to orbit Saturn. As a result, these moons always keep the same side facing the planet. Astronomers call this a captured rotation. However, the outermost of Saturn's moons, Phoebe, does not have a captured rotation.

### The Shepherd Moons

Of Saturn's smallest moons, some of the most interesting are those that orbit close to the rings. Two moons, Pan and Atlas, orbit close to the edge of the A ring. Like many of Saturn's small moons, they are rocky lumps with an irregular shape. Atlas is about 19 miles (30 km) across, and Pan is even smaller. Just beyond Atlas is another pair of somewhat larger moons—Prometheus, which has a diameter of about 62 miles (100 km), and Pandora, which is about 55 miles (90 km) across. Prometheus orbits just inside the F ring, and Pandora travels just outside of it.

Astronomers believe that these tiny moons play an important part in keeping particles within the nearby rings. They are called shepherd moons because of the way they seem to "herd" the ring particles. As these moons travel around either side of the ring, their gravity helps pull back any particles that stray outside the rings.

This montage of photographs shows Saturn and some of its largest moons. Clockwise from top right, the moons are Titan, Mimas, Tethys, Dione, Enceladus, and Rhea.

### In the Same Orbit

The next small moons beyond Pandora are called Epimetheus and Janus. Both moons are less than 125 miles (200 km) across. These moons are so close together that they seem to travel in the same orbit. Astronomers call them co-orbitals. In fact, they do travel in slightly different orbits at slightly different speeds. The inner moon, Epimetheus, travels slightly faster than the outer moon, Janus. Every four years, when Epimetheus catches up with Janus, the gravitational forces between the two moons causes them to swap orbits. They swap back again the next time they meet.

A number of Saturn's small moons share orbits with larger moons. Telesto and Calypso share an orbit with a larger moon, Tethys. And Helene shares an orbit with another larger moon, Dione. The three small co-orbitals are only about 19 miles (30 km) in diameter.

Below are some of the tiny moons that orbit Saturn. Some are less than 20 miles (30 km) across.

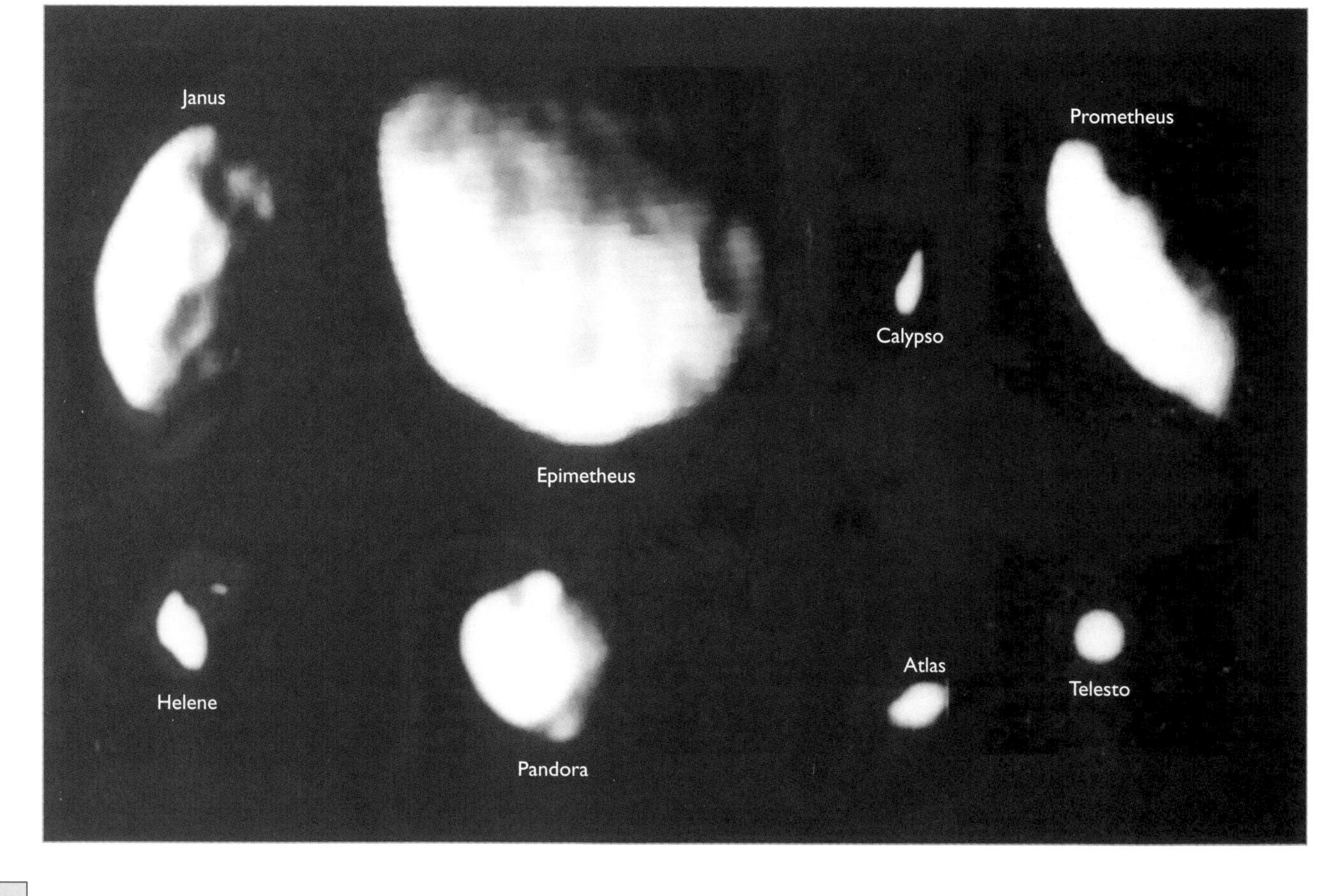

## Small and Distant

Two small moons orbit at a great distance from Saturn. Hyperion, with a diameter of 180 miles (290 km), is nearly 1.5 million miles (2.4 million km) from Saturn. About the same size as Hyperion, Phoebe is much more remote—nearly 13 million miles (21 million km) from Saturn. This moon takes over 550 days to circle around the planet. That's longer than Earth takes to circle the Sun!

Phoebe is the smallest of Saturn's moons that can be seen through a telescope. It is fairly dark with an irregular shape. Phoebe travels in its orbit in the opposite direction of Saturn's other moons. This suggests that it was not originally part of Saturn's family but was a small body that once orbited the Sun. It was probably captured by the planet's gravity long after the other moons began orbiting Saturn.

## Iceball Moons

Saturn's larger moons are spherical, and all but Titan are covered with ice. Five of these round icy moons orbit relatively close to Saturn, beyond the ring system. In order going out from the planet, they are Mimas, Enceladus, Tethys, Dione, and Rhea. The sixth iceball moon, Iapetus, orbits between Hyperion and Phoebe.

All these moons seem to be made up of a mixture of water ice and rock. Beneath their icy surface, some of the larger moons may have a rocky core.

| Moon | Diameter | | Spins on axis in |
|---|---|---|---|
| | miles | km | |
| Mimas | 244 | 392 | 0.9 days |
| Enceladus | 310 | 500 | 1.4 days |
| Tethys | 660 | 1,060 | 1.9 days |
| Dione | 695 | 1,120 | 2.7 days |
| Rhea | 950 | 1,530 | 4.5 days |
| Iapetus | 905 | 1,460 | 79.3 days |

Like Saturn's other large moons, Dione is covered with craters. The largest one in this picture is about 60 miles (100 km) across.

## Mimas

Mimas is the closest of Saturn's moons that can be seen through telescopes on Earth. English astronomer William Herschel discovered it in 1789. Its icy surface is covered with craters, or large pits, and there are large cracks in the crust, which astronomers call chasma. The most striking marking on Mimas is the crater Herschel, named after the moon's discoverer. This crater is 80 miles (130 km) wide, which is almost one-third of the moon's entire diameter. Certain parts of the crater's floor are 6 miles (10 km) deep, and its walls are 3 miles (5 km) high. It was originally named Arthur after the legendary English king. Other surface features are named after characters from the Arthurian legend, such as the crater Merlin and the valley Camelot.

**Above:** Mimas is one of the most heavily cratered of Saturn's moons.

## Enceladus

Although Enceladus is not much bigger than Mimas, its surface is quite different. Instead of being completely covered in craters, this moon has several different kinds of surfaces. Some parts of Enceladus are cratered, some are covered by long grooves and cracks, and some parts are smooth, with few features at all. These smooth areas are the most mysterious. Astronomers think they were probably created by ice that melted and froze over again, but no one knows exactly what caused the ice to melt.

Enceladus has smooth, cratered, and cracked areas on its surface.

Enceladus is the most reflective body in the solar system. It reflects nearly all the light that reaches it. In contrast, our own Moon reflects only about 7 percent of the light it receives.

## Tethys

Astronomers think that this moon must be made up almost completely of water ice, with hardly any rocky material at all. Like Mimas, it is heavily cratered. One crater, Odysseus, is more than 250 miles (400 km) across. It is so big that Mimas could fit into it. Tethys's other remarkable feature is a huge fault, or crack, in its crust, named Ithaca Chasma. In some places the fault is several miles deep and more than 60 miles (100 km) wide. It stretches for more than 1,200 miles (2,000 km). The only other feature we know like it in the solar system is Valles Marineris on Mars.

Tethys is covered with large and small craters.

## Dione

Although it is nearly the same size as Tethys, Dione is much heavier, which means it contains more rocky material. Dione is somewhat darker on one side than the other and has bright wispy streaks. These streaks might be caused by fresh icy material that has forced its way to the surface through cracks in the crust. There are also several large craters on Dione, up to 150 miles (240 km) across and with central mountain peaks. Most of Dione's craters are small.

In this photo of Saturn and two of its moons, Tethys and Dione, Tethys is making a shadow on the planet, just below the rings on the far right.

## Rhea

Rhea is Saturn's second largest moon. It appears to be made up of equal amounts of ice and rock and probably has a rocky core. Although Rhea is somewhat larger than Dione, it looks very similar. In particular, one side is slightly darker than the other, and it has similar bright wispy markings. Also like Dione, Rhea is heavily cratered, with cracks in some parts of its surface.

## Iapetus

Only slightly smaller than Rhea, Iapetus lies at a great distance from Saturn. At about 2.2 million miles (3.5 million km) away, it orbits among Saturn's most distant moons. Iapetus is one of the more mysterious of Saturn's moons. Even from Earth, astronomers notice that at times the moon is very bright, while at other times it almost disappears. This is because one side of the moon is much brighter than the other. The bright side seems to have an icy, cratered surface, like most of Saturn's moons. But the other side is coated with a dark material that does not reflect much light. Astronomers are not sure what the dark coating is. It may be the result of inner materials that have welled up to the surface.

Unlike Saturn's large moons, small moons like Hyperion have an irregular shape.

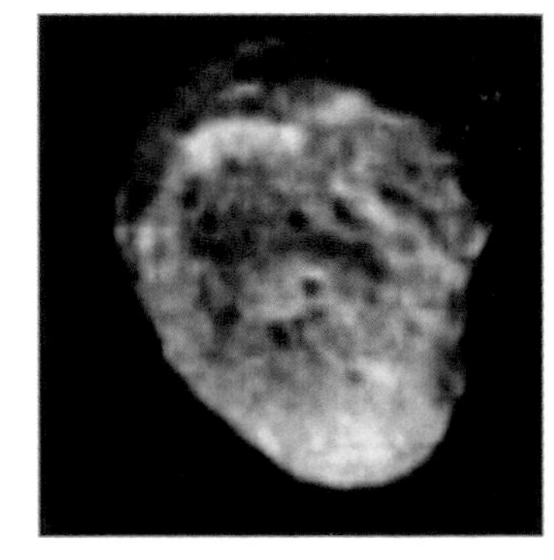

Ganymede

Earth's Moon

Titan

# Under Titan's Clouds

**Titan is Saturn's largest moon and the second largest moon in the solar system. Its most unusual feature is its thick atmosphere.**

Titan is nearly as large as Jupiter's Ganymede, the largest moon in the solar system. Both moons are much bigger than our Moon.

When Dutch astronomer Christiaan Huygens discovered Titan in 1655, other moons around Saturn had not been discovered. Because Titan is so large, it is easy to spot, even with binoculars. With a diameter of 3,200 miles (5,150 km), it is larger than the planet Mercury but smaller than Mars. Among the moons in the solar system, only Jupiter's Ganymede is bigger. Titan orbits Saturn at an average distance of about 750,000 miles (1,200,000 km) and lies relatively close to Hyperion, one of Saturn's outermost moons. Titan takes about 16 days to circle Saturn.

## Titan's Atmosphere

Titan's most remarkable feature is its thick atmosphere—no other moon in the solar system has one like it. The orange atmosphere, with its deep layers of clouds and a smog-like haze, prevents us from seeing the moon's surface.

Nitrogen, the main gas found on Earth, makes up most of Titan's atmosphere. Other gases, such as methane, hydrogen, and argon, are also present, but only in very small amounts. Because Titan contains a mixture of gases similar to some of those found on Earth, scientists have wondered if Titan could support life. But as far as we know, temperatures on Titan are much too low for any forms of life to exist.

An artist's impression of Titan's surface, with Saturn in the distance

*Pioneer Saturn* turned its instruments and cameras on Saturn in September 1979. It provided us with much of our information about the planet.

# Missions to Saturn

***Pioneer* and *Voyager* space probes have sent back huge amounts of new information about the mysterious ringed planet.**

In April 1973, NASA (National Aeronautics and Space Administration) launched the *Pioneer 11* space probe toward Jupiter. It was following in the footsteps of the identical *Pioneer 10*, launched the previous year. After its encounter with Jupiter in December 1973, *Pioneer 10* began heading out of the solar system. But *Pioneer 11*, after encountering Jupiter in December 1974, headed for an encounter with Saturn. It was renamed *Pioneer Saturn*.

To reach Saturn from Jupiter, the spacecraft had to travel to the other side of the solar system. The journey took nearly five years. In September 1979, *Pioneer Saturn* made its closest approach to Saturn when it came within 13,000 miles (21,000 km) of the planet.

*Pioneer Saturn* sent back the most detailed and close-up pictures of the planet ever seen. It allowed scientists to discover the F ring just outside the A ring, and a new moon, which mission astronomers nicknamed Pioneer Rock.

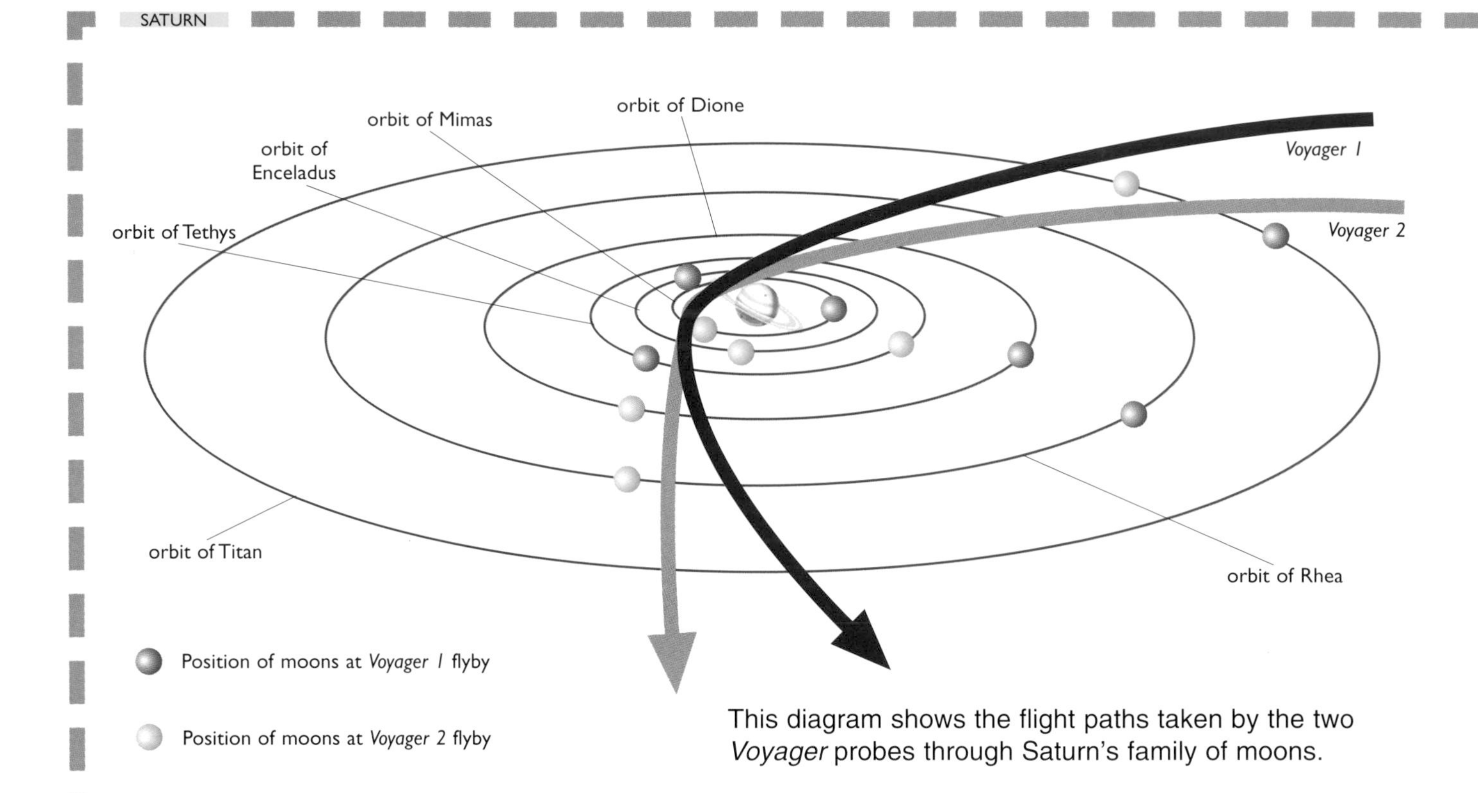


This diagram shows the flight paths taken by the two *Voyager* probes through Saturn's family of moons.

## The *Voyager* Invasion

*Pioneer Saturn* was a pathfinding mission for the next probes to make their way to the ringed planet. They were *Voyager 1* and *Voyager 2*. Like *Pioneer Saturn*, the *Voyagers* visited Jupiter first, where they made astounding discoveries. Mission scientists were hoping for similar success when the probes encountered Saturn. They were not disappointed.

In September 1997, the *Voyagers* celebrated their 20th anniversary in space. *Voyager 1* was over 6 billion miles (10 billion km) away, and *Voyager 2* was nearly 5 billion miles (8 billion km) away.

## Revealing Images

During October 1980, as *Voyager 1* homed in on its target, it began revealing Saturn as never before. Clouds and spots showed up in the atmosphere. New moons were discovered. The rings proved to be made up of thousands of separate ringlets. And the mysterious spokes were observed. The probe reached the ringed planet in November 1980.

Day after day until early December, images of Saturn flooded into mission control at the Jet Propulsion Laboratory in Pasadena, California. Each day they revealed something new, such as a new moon, ringlets in the ring gaps, and craters on Saturn's moons.

### Another Voyage

By mid-December 1980, *Voyager 1*'s main mission of investigating Saturn was over, and it headed off into outer space. Meanwhile, *Voyager 2* was still a long way from Saturn. It had taken a different path through space toward its target. This was planned so that its course would take it not only to Saturn, but on to more distant planets—Uranus and Neptune. So *Voyager 2* did not close in on Saturn until the summer of 1981. It repeated *Voyager 1*'s success and returned floods of data and tens of thousands of exciting images.

NASA had aimed *Voyager 2* with great precision. When it made its closest approach to Saturn, on August 25, 1981, it was less than 30 miles (48 km) off target and just 3 seconds ahead of schedule. This was amazingly accurate after a four-year journey of over 1 billion miles (1.6 billion km).

Pictures taken by the *Voyagers* are stunningly beautiful. This one clearly shows the bands in Saturn's atmosphere.

**Inset:** This close-up view of Saturn shows the complicated patterns of wind flow in the planet's atmosphere.

## Cassini's Encounter

Successful though they were, the *Voyager* probes paid only a fleeting visit to Saturn. That is why American and European space scientists decided to launch a follow-up mission called *Cassini-Huygens*.

The 5½-ton *Cassini-Huygens* spacecraft blasted off from Cape Canaveral in October 1997. Ahead was a seven-year journey that aimed to put the probe into orbit around Saturn in the early part of the 21st century. On its way, the probe was directed to loop past Venus, Earth, and Jupiter and use those planets' gravity to gain extra speed.

The probe has two parts. NASA's *Cassini* craft is the main spacecraft, designed to orbit Saturn for four years. It carries cameras and instruments to explore Saturn's atmosphere, its rings, and its many moons. The European Space Agency's *Huygens* probe is a landing craft, designed to parachute into Titan's atmosphere and touch down on its surface. Huygens carries instruments to analyze the gases in Titan's unique atmosphere, measure the strength of its winds, and take the first pictures of its surface.

In this artist's impression of Titan, the *Huygens* probe parachutes down to a methane ocean with icy shores. Scientists are eager to learn about Titan's unknown surface.

Scientists think that Titan's atmosphere might be like Earth's original atmosphere. They hope that the information *Huygens* sends back will help them understand more about what Earth was like in its early formation.

# Glossary

**atmosphere:** the layer of gases around a planet or other heavenly body

**aurora:** a glow in a planet's atmosphere produced by particles from the Sun that enter the planet's magnetic field before reaching its atmosphere

**axis:** an imaginary line passing through a planet from its north to its south pole

**belt:** a dark-colored band on Saturn's atmosphere that runs parallel to the equator

**captured rotation:** when a moon rotates on its axis in the same amount of time that it takes to orbit its planet once

**chasma:** a crack, or valley, on the surface of a planet or moon.

**co-orbital:** a moon that travels in the same orbit as another moon

**core:** the center part of a planet, moon, or other heavenly body

**crater:** a pit on the surface of a planet, moon, or other body

**crust:** the hard outer layer of a rocky or icy planet or moon

**equator:** an imaginary line around a planet midway between its north and south poles

**gas giant:** a large planet that is made up mainly of gas; Jupiter, Saturn, Uranus, and Neptune are the gas giants in our solar system

**gravity:** the attraction, or pull, that every heavenly body has on objects on or near it

**jet stream:** a fast-moving air current in a planet's atmosphere

**magnetosphere:** a magnetic region in space around a planet

**orbit:** the path in space of one heavenly body around another, such as Saturn around the Sun

**planet:** a large body that orbits the Sun

**probe:** a spacecraft that travels from Earth to explore bodies in the solar system

**ringlet:** a very narrow ring

**ring system:** a set of rings around a giant planet, made up of fine rocky and icy particles

**shepherd moon:** a tiny moon located near a planet's ring that may help keep the ring particles in place

**solar system:** the Sun and all the bodies that circle around it, including Saturn and the other planets

**solar wind:** a fast-moving stream of particles that travels outward from the Sun

**spoke:** dark lines fanning out on Saturn's B ring that are similar to the spokes of a wheel

**zone:** a light-colored band on Saturn's atmosphere that runs parallel to the equator

# Index

AWML
DISCARD